ProMgmt.®

Management by Menu

THIRD EDITION

Student Workbook

National Restaurant Association
EDUCATIONAL FOUNDATION

JOHN WILEY & SONS, INC.
New York • Chichester • Weinheim • Brisbane • Singapore • Toronto

ProMgmt. is a registered trademark of the National Restaurant Association Educational Foundation.

This student workbook is designed to be used with the textbook *Management by Menu, Third Edition* by Lendal H. Kotschevar and Marcel R. Escoffier.

This book is printed on acid-free paper.

Published by John Wiley & Sons, Inc.

Published simultaneously in Canada.

Library of Congress Cataloging-in-Publication Data:

ISBN: 0-471-41320-8

Printed in the United States of America.

10 9 8 7 6 5 4 3 2

Contents

Introduction 1

Lesson 1 5
Overview of the Foodservice Industry
Text Chapters 1 and 2

Lesson 2 11
Developing the Menu and Cost Controls
Text Chapters 3–5

Lesson 3 19
Menu Pricing, Mechanics, and Analysis
Text Chapters 6–8

Lesson 4 27
The Liquor Menu and Planning a Healthy Menu
Text Chapters 9 and 10

Lesson 5 33
The Menu and Purchasing, Production, and Service
Text Chapters 11 and 12

Lesson 6 39
Computers and Finances in Menu Planning
Text Chapters 13 and 14

Study Outline 45

Practice Test 65

Practice Test Answers 75

Introduction

The *Management by Menu, Third Edition* course will give future foodservice managers a clear picture of the important role menu planning plays within an operation. This course begins with an overview of the foodservice industry and its history. It covers topics ranging from menu development and pricing, to production and service.

Students will benefit from this course because understanding menus is crucial to the success of any foodservice operation. A menu is more than just the sum of its parts: it is a planning tool, a source of operational information, and the primary merchandising method for reaching patrons. Good menu management is important to the success of all managers in foodservice operations.

How to Earn a ProMgmt℠ Certificate of Course Completion

To earn a ProMgmt. Certificate of Course Completion, a student must complete all student workbook exercises and receive a passing score on the final examination.

To apply for the ProMgmt. Certificate of Course Completion, complete the student registration form located on the inside back cover of this workbook and give it to your instructor, who will then forward it to the National Restaurant Association Educational Foundation.

Each student registered with the Educational Foundation will receive a student number. Please make a record of it; this number will identify you during your present and future coursework with the Educational Foundation.

ProMgmt. certificate requirements are administered exclusively through colleges and other educational institutions that offer ProMgmt. courses and examinations.

If you are not currently enrolled in a ProMgmt. course and would like to earn a ProMgmt. certificate, please contact your local educational institution to see if they are willing to administer the ProMgmt. certificate requirements for non-enrolled students. You can also visit **www.edfound.org** for a list of ProMgmt. Partner schools. ProMgmt. Partner schools offer seven or more courses that include administration of the ProMgmt. certificate requirements.

The Educational Foundation leaves it to the discretion of each educational institution offering ProMgmt. courses to decide whether or not that institution will administer the ProMgmt. certificate requirements to non-enrolled students. If an institution does administer ProMgmt. certificate requirements to non-enrolled students, that institution may charge an additional fee, of an amount determined by that institution, for the administration of the ProMgmt. certificate requirements.

Course Materials

This course consists of the text, *Management by Menu, Third Edition,* by Lendal H. Kotschevar, Ph.D., FMP and Marcel R. Escoffier, the student workbook, and a final examination. The examination is the final section of your course and is sent to an instructor for administration, then returned to the Educational Foundation for grading.

Each lesson consists of:

- Student objectives
- Reading assignment
- Chapter exercises

At the end of the Workbook you will find:

- A study outline of the textbook
- A glossary (when the textbook does not have one)
- An 80-question practice test
- Answers to the practice test

The objectives indicate what you can expect to learn from the course, and are designed to help you organize your studying and concentrate on important topics and explanations. Refer to the objectives frequently to make sure you are meeting them.

The exercises help you check how well you've learned the concepts in each chapter. These will be graded by your instructor.

An 80-question practice test appears at the end of the workbook. All the questions are multiple-choice and have four possible answers. Circle the best answer to each question, as in this example:

Who was the first president of the United States?

A. Thomas Jefferson
(B.) George Washington
C. Benjamin Franklin
D. John Adams

Answers to the practice test follow in the workbook so that you may grade your own work.

The Final Exam

All examinations may first be graded by your instructor and then officially graded again by the Educational Foundation. If you do not receive a passing grade on the examination, you may request a retest. A retest fee will be charged for the second examination.

Study Tips

Since you have already demonstrated an interest in furthering your foodservice education by registering for this Educational Foundation course, you know that your next step is study preparation. We have included some specific study pointers which you may find useful.

- Build studying into your routine. If you hold a full-time job, you need to take a realistic approach to studying. Set aside a specific time and place to study, and stick to your routine as closely as possible. Your study area should have room for your course materials and any other necessary study aids. If possible, your area should be away from family traffic.
- Discuss with family members your study goals and your need for a quiet place and private time to work. They may want to help you draw up a study schedule that will be satisfactory to everyone.
- Keep a study log. You can record what lesson was worked on, a list of topics studied, the time you put in, and the dates you sent your exercises to your instructor for grading.
- Work at your own pace, but move ahead steadily. The following tips should help you get the most value from your lessons.
    1. Look over the objectives carefully. They list what you are expected to know for the examination.
    2. Read the chapters carefully, and don't hesitate to mark your text—it will help you later. Mark passages that seem especially important and those that seem difficult, as you may want to reread these later.

3. Try to read an entire chapter at a time. Even though more than one chapter may be assigned in a lesson, you may find you can carefully read only one chapter in a sitting.
4. When you have finished reading the chapter, go back and check the highlights and any notes you have made. These will help you review for the examination.

Reviewing for the Final Exam

When you have completed the final exercise and practice test, you will have several items to use for your examination review. If you have highlighted important points in the textbook, you can review them. If you have made notes in the margins, check them to be sure you have answered any questions that arose when you read the material. Reread certain sections if necessary. Finally, you should go over your exercises.

The ProMgmt. Program

The National Restaurant Association Educational Foundation's ProMgmt. Program is designed to provide foodservice students and professionals with a solid foundation of practical knowledge and information. Each course focuses on a specific management area. For more information on the program, please contact the National Restaurant Association Educational Foundation at 800.765.2122 (312.715.1010 in Chicagoland) or visit our web site at **www.edfound.org.**

Lesson 1

OVERVIEW OF THE FOODSERVICE INDUSTRY

Student Objectives

After completing this lesson, you should be able to:

- Discuss the development of foodservice during ancient times and the European Middle Ages.
- Describe the development of *haute cuisine* and early restaurants in France.
- Discuss the contributions of English, Russian, and Italian cuisines to classic cooking.
- Evaluate the impact of the Industrial Revolution and modern science on the foodservice industry.
- Recognize the importance of Marie Antoine Carême, Georges Auguste Escoffier, and César Ritz to the "Golden Age" of cuisine.
- Examine the early years of food service in the United States, and the rapid industry growth following World War II.
- Explain the division of income and costs and how to determine profitability.
- Describe the following segments of the foodservice industry: commercial feeding, institutional feeding, transportation feeding, health services feeding, clubs, military feeding, and central commissaries.
- Discuss how the economy, society, labor force, employee turnover, eating habits, government regulations, industry factors, and technology are likely to affect the future of the industry.

Reading Assignment

Now read Chapters 1 and 2 in the text. Use this information to answer the questions and activities in Exercises 1 and 2.

Chapter 1 Exercise

1. Briefly summarize the contribution(s) each person below made to the foodservice industry.

   a. Apicius ____

   ____

   b. Catherine de Medici ____

   ____

   c. Boulanger ____

   ____

   d. Carême ____

   ____

   e. Escoffier ____

   ____

2. How did the rise of the middle class during the Industrial Revolution impact the foodservice industry?

   ____

   ____

   ____

   ____

   ____

3. How have modern scientific developments impacted the foodservice industry?

   ____

   ____

   ____

   ____

   ____

Chapter 2 Exercise

1. Match each numbered foodservice operation with the lettered category that describes it. Letters will be used more than once.

____(1) Prison
____(2) Social caterer
____(3) Family-style restaurant
____(4) Nursing home
____(5) Vending service
____(6) Elementary school
____(7) Contract food service
____(8) Recreational food service

a. Commercial
b. Institutional

2. Match each numbered foodservice operation with its lettered definition.

____(1) Food contractor
____(2) Social caterer
____(3) Retail host
____(4) Central commissary
____(5) Quick-service restaurant

a. Large kitchen that can mass-produce foods quickly for shipment to satellite units
b. Restaurant that provides quickly prepared and served items
c. Organization that contracts to produce and serve food for other businesses
d. Operation that prepares food and beverages in a central kitchen for service elsewhere
e. Establishment, such as a supermarket, department store, or gas station, that offers food in conjunction with their other services

3. Describe how changes in family composition affect the foodservice industry.

____

____

____

____

____

4. Which demographic group's presence in the labor force is expected to grow at the slowest pace in the next 20 years?

____

____

5. How is competition among foodservice operators expected to change around the year 2000?

____

____

____

____

____

Lesson 2

Developing the Menu and Cost Controls

Student Objectives

After completing this lesson, you should be able to:

- Explain how a menu functions as both a sales tool and a production sheet.
- Use a mission statement to plan organizational objectives, specific strategic plans, and operating budgets.
- Discuss who prepares the menu using what tools.
- Compare the different types of menus: à la carte, table d'hôte, du jour, limited, cycle, and California.
- Describe how meal plans influence the menu and menu organization.
- Categorize typical menu offerings and determine the number of items to include on a menu.
- Compare menu items for different meals and special occasions.
- Discuss menus designed to meet the special needs of children, teens, and senior citizens.
- Examine how the physical factors of the facility, labor considerations, guest expectations, and the appearance and variety of food influence menu planning.
- Evaluate food preferences.
- Describe obtaining and monitoring operation and labor costs.
- Discuss institutional and commercial cost control, and controlling food, labor, and budget costs.

Reading Assignment

Now read Chapters 3–5 in the text. Use this information to answer the questions and activities in Exercises 3–5.

Chapter 3 Exercise

1. While a menu serves as a tool for customers—informing them of what is available and at what price—describe how a menu serves as a tool for foodservice managers.

   ___

   ___

2. Explain the difference between a strategic plan and a tactical plan.

   ___

   ___

3. How is market research helpful in developing a menu?

   ___

   ___

4. List two methods of conducting market research for food service.

   - ___
   - ___

5. Explain the difference between an à la carte menu and a table d'hôte menu.

   ___

   ___

6. Explain the difference between a cycle menu and a California menu.

   ___

   ___

7. What are the two key things a patron decides when selecting a meal from a menu?

   - ___
   - ___

8. Within the entree category, it is typical to split the menu items into four categories. What are these categories?

- ______________________________
- ______________________________
- ______________________________
- ______________________________

9. A breakfast menu should typically offer eight categories of items. What are these eight categories?

- ______________________________
- ______________________________
- ______________________________
- ______________________________
- ______________________________
- ______________________________
- ______________________________
- ______________________________

10. In what primary way does a reception menu differ from a tea menu?

____

____

Chapter 4 Exercise

1. Explain how the equipment and facilities available in a foodservice establishment influence menu planning.

____

____

2. Explain how labor considerations influence menu planning.

____

____

3. What is the difference between a physiological need and a perceived need?

   Physiological need:_______________

   ____

   ____

   Perceived need:_______________

   ____

   ____

4. Why should a menu offer a sufficient variety of items to its customers?

____

____

5. Why should a menu planner avoid offering too much variety to his or her patrons?

____

____

6. Explain why appearance is a key consideration when attempting to create appealing food.

____

____

7. Explain the difference between the texture and the consistency of food.

Texture: ______________________________________

Consistency: __________________________________

8. Why should a menu present exact descriptions of food and use simple language?

____

____

9. How do the federal government's Standards of Identity influence the wording of a menu?

____

____

Chapter 5 Exercise

1. Using the following information, calculate the total food cost. Beginning inventory = $60,000; purchases, transportation, delivery and other charges = $100,000; ending inventory = $70,000; and total sales = $300,000.

2. Use the information in question 1 to calculate the food cost percentage.

3. If a hospital food service is on a set budget of $25,500 a month, 32% of the budget is used to pay labor, and 16% is used to pay other expenses, how much money is available for total food cost allocation?

4. What is the spindle method of compiling food cost?

   ____

   ____

5. Calculate portion cost using the following information: one portion of ice cream = 4 oz; you have purchased 5 quarts of ice cream for $12.50. (There are 32 oz in 1 quart.)

6. Explain the two purposes of a yield test.

   - ____
   - ____

7. What two costs are combined to obtain the prime cost of an item?
   - ______________________________
   - ______________________________

8. Typically, food and labor costs should not exceed what percent? ____________

9. What are two things that can be done to achieve correct portion size?
   - ______________________________
   - ______________________________

10. If a noncommercial foodservice operation produces 20,000 meals per month and wants to produce 5 meals for every hour of labor used, how many labor hours would the operation require per month?

11. What is the difference between a job description and a job specification?

    Job description: ______________________________

    ____

    Job specification: ______________________________

    ____

Lesson 3

Menu Pricing, Mechanics, and Analysis

Student Objectives

After completing this lesson, you should be able to:

- Describe how customers' value perception, pricing psychology, and market research influence menu prices.
- Describe pricing methods and how to evaluate them.
- Discuss special cases, including pricing for nonprofit organizations, pricing employee meals, and changing menu prices.
- Explain the importance of menu presentation and format.
- Consider factors such as menu type size and spacing, the weight and texture of paper, item emphasis, menu color, and cover design.
- Discuss printing methods.
- Analyze menus using item counts, subjective evaluation, and popularity indexes.
- Use menu factor analysis, the Hurst method, the break-even method, and other methods to perform menu analysis.

Reading Assignment

Now read Chapters 6–8 in the text. Use this information to answer the questions and activities in Exercises 6–8.

Chapter 6 Exercise

1. If a menu item costs $1.53 and the operation sells it for $6.95, what is the markup on the item?

2. What is value perception?

   ____

   ____

3. Why do many menu planners avoid whole number pricing and opt, instead, to shade numbers just below whole numbers?

   ____

   ____

4. What are the two basic laws of economics affecting every operation?

   - ____
   - ____

5. Use the derived food cost method to calculate a menu price of an item based on the following information: An item has a total food cost of $4.90, and the operator wants a food cost percentage of 35%.

6. Use the gross profit pricing method to calculate the average gross profit per guest using the following information: the establishment has total sales of $930,000; total food cost of $330,000; and total number of guests served is 120,000.

7. If a menu item costs $1.95 and the average gross profit per guest is $5.00, what is an appropriate selling price for the item?

8. What are two of the three non-cost pricing methods?

   - ____________________
   - ____________________

9. What are two of the four pricing pitfalls a menu planner should avoid?

   - ____________________
   - ____________________

Chapter 7 Exercise

1. What are the two main functions of a successful menu?
   - ______________________________
   - ______________________________

2. What, typically, is on the front of a menu?

3. What are the four types of print?
   -
   -
   -
   -

4. How many points are in an inch?

5. What point type do menu planners typically use for headings?

6. What two things can be done to make a menu item stand out from other items?
   -
   -

7. What is a print's weight?

8. Typically, what percentage of a menu has print on it?

9. Do the majority of menu readers prefer a single-column page or a double-column page?

10. Why is color a vital part of menu design?

11. What is cover stock?

12. What is the most prominent part of a two-page menu?

Chapter 8 Exercise

1. List four of the seven methods of menu analysis.
   - ______________________________
   - ______________________________
   - ______________________________
   - ______________________________

2. If an establishment sells 325 pizzas, 350 hamburgers, 225 tacos, 525 orders of ribs, and 575 orders of French fries, what is the popularity index of each item?

   Pizza: ______________________________

   Hamburgers: ______________________________

   Tacos: ______________________________

   Ribs: ______________________________

   French fries: ______________________________

3. Menu factor analysis develops factors to indicate how an item is doing in four specific areas. List these areas.
   - ______________________________
   - ______________________________
   - ______________________________
   - ______________________________

4. Calculate the goal value standard using the following information: average food cost percentage is 30%; average number (volume) sold is 25; average selling price is $6.00; variable food cost is 45%.

5. When conducting a break-even analysis, a menu planner must calculate a break-even point. What three things must be known in order to calculate this point?

- ______________________________
- ______________________________
- ______________________________

Lesson 4

THE LIQUOR MENU AND PLANNING A HEALTHY MENU

Student Objectives

After completing this lesson, you should be able to:

- Discuss ways to present a liquor menu.
- Describe typical elements of wine and spirits lists, and how to merchandise items.
- Characterize spirits, beers, ales, liqueurs, and cordials.
- Describe proper wine and beverage service.
- Discuss considerations involved in liquor pricing.
- Explain controls used in purchasing, receiving, issuing, production, and service of beverages.
- Consider the criticism that the foodservice industry contributes to Americans' poor nutrition, and customers' and operators' responsibilities for nutritious eating.
- Identify and describe the building blocks of nutrition: carbohydrates, proteins, fats, vitamins, minerals, and water.
- Discuss how to retain nutrients in quantity food preparation.
- Describe customers' major health concerns regarding food: that it be low in calories, fat, cholesterol, and sodium.

Reading Assignment

Now read Chapters 9 and 10 in the text. Use this information to answer the questions and activities in Exercises 9 and 10.

Chapter 9 Exercise

1. What are the five categories of liquor typically found on a liquor menu?

   - ______________________________
   - ______________________________
   - ______________________________
   - ______________________________
   - ______________________________

2. What are the three broadest categories for wine, based on its color?

   - ______________________________
   - ______________________________
   - ______________________________

3. Are white wines traditionally served chilled or unchilled? ______________

4. Match each item on the left with the category on the right that describes it. Letters will be used more than once.

____(1) Brandy	a. Wine
____(2) Chablis	b. Spirit
____(3) Vodka	c. Liqueur/Cordial
____(4) Crême de menthe	
____(5) Rum	
____(6) Chardonnay	
____(7) Burgundy	
____(8) Whiskey	

5. Which of the following typically has the highest alcohol content—wines, beers/ales, or spirits? ______________________

6. What are a sommelier's duties?

____

____

7. What are the two ways liquor pricing typically differs from food pricing?

- ____
- ____

8. List three methods commonly used to reconcile liquor sales.

- ____
- ____
- ____

Chapter 10 Exercise

1. What is the primary drawback to using natural foods?

____

____

2. What are three of the five criteria a menu planner can follow in order to provide patrons with healthier food?

- ____
- ____
- ____

3. The United States Department of Agriculture has developed a Food Pyramid that indicates how many servings of each food group should be consumed daily. Match each numbered food item with its lettered correct number of servings. Letters will be used more than once.

____(1) Carrots
____(2) Apples
____(3) Poultry
____(4) Potatoes
____(5) Cereal
____(6) Eggs
____(7) Bread
____(8) Milk
____(9) Rice
____(10) Butter
____(11) Kidney beans
____(12) Refined sugar

a. 2–5 servings
b. 6–11 servings
c. Sparingly

4. Match each food item on the left with the category on the right that describes it. Letters will be used more than once.

____(1) Egg
____(2) Potato
____(3) Fish
____(4) Bread
____(5) Cheese
____(6) Macaroni

a. Carbohydrate
b. Protein

5. Match each vitamin on the left with the food on the right that contains a large percentage of it. Letters will be used more than once.

____(1) Vitamin A	a. Cereal
____(2) Vitamin B complex	b. Leafy, green vegetable
____(3) Vitamin C	c. Citrus fruit
____(4) Vitamin E	
____(5) Vitamin K	

6. List three of the five things cooks can do to preserve the nutritional quality of food.

- ____
- ____
- ____

7. In order to provide patrons with healthy food choices, the menu should offer foods low in what three things?

- ____
- ____
- ____

Lesson 5

THE MENU AND PURCHASING, PRODUCTION, AND SERVICE

Student Objectives

After completing this lesson, you should be able to:

- Explain the purchasing function.
- Describe how the purchaser determines need, searches the market, writes specifications, and makes a value analysis.
- Explain controls used in receiving, storage, inventory, and issuing of items.
- Analyze why production determines the success of a menu.
- Identify the goals of service.
- Describe the tasks involved in service mise en place, greeting customers, serving food, and delivering the check.
- Characterize different types of service: counter, cafeteria, buffet, and seated.
- Compare and contrast American, French, English, and Russian service.
- Explain the practice of tipping.
- Discuss the most efficient way to control guest checks.
- Describe how servers should present a guest check and receive payment.

Reading Assignment

Now read Chapters 11 and 12 in the text. Use this information to answer the questions and activities in Exercises 11 and 12.

Chapter 11 Exercise

1. What are the six steps of purchasing?

   (1) ______________________________

   (2) ______________________________

   (3) ______________________________

   (4) ______________________________

   (5) ______________________________

   (6) ______________________________

2. What is the job of the purchaser?

   ____

   ____

3. What are the two factors essential to value in any menu?

   - ______________________________
   - ______________________________

4. What are the three aspects of searching the market?

   - ______________________________
   - ______________________________
   - ______________________________

5. List four of the six pieces of information a specification should include.

   - ______________________________
   - ______________________________
   - ______________________________
   - ______________________________

6. What is a Universal Product Code?

____

____

7. What is a standard of identity?

____

____

8. What are the three aspects of purchasing control?

- ____
- ____
- ____

9. What is value analysis?

____

____

10. List three documents that help managers control labor.

- ____
- ____
- ____

Chapter 12 Exercise

1. What is the concept of mise en place? Why is it vital to successful customer service?

   ____

   ____

2. A good serving system enables what four things to be done properly?
   - ______________________________
   - ______________________________
   - ______________________________
   - ______________________________

3. Servers should be trained in what six areas?
   - ______________________________
   - ______________________________
   - ______________________________
   - ______________________________
   - ______________________________
   - ______________________________

4. In what direction does table service typically occur?

   ____

   ____

5. What are the three main types of service?
   - ______________________________
   - ______________________________
   - ______________________________

6. What is scramble cafeteria service?

7. What is the main advantage of buffet service?

8. What is the main disadvantage of buffet service?

9. Why is French table service slower and more expensive than American table service?

10. Where should servers place the check if they are not sure which guest is paying?

11. Where should servers place the check when they know which guest is paying?

12. How does time stamping help guard against floating checks?

Lesson 6

COMPUTERS AND FINANCES IN MENU PLANNING

Student Objectives

After completing this lesson, you should be able to:

- Describe how computers are used in front-of-the-house and back-of-the-house applications.
- Identify how the computer is used in management decision making.
- Explain how capital investment supports the operation.
- Analyze how food service may be perceived as a high-risk business.
- Discuss two factors that largely determine the success of the operation: location and market.
- Discuss the benefits and drawbacks of owning, renting, and leasing an establishment.
- Identify the parts of a feasibility study and why it is important to conduct one.
- Measure profitability and return on investment to determine an operation's success.
- Discuss the liquidity of a commercial operation and the efficiency of a noncommercial operation.

Reading Assignment

Now read Chapters 13 and 14 in the text. Use this information to answer the questions and activities in Exercises 13 and 14.

Chapter 13 Exercise

1. Match each activity on the left with the part of the business on the right to which it pertains. Letters will be used more than once.

____(1) Sales forecasting
____(2) Guest sales transactions
____(3) Accounting
____(4) Sales analysis
____(5) Vendor transactions
____(6) Server productivity
____(7) Budget analysis
____(8) Financial reporting

a. Front of the house
b. Back of the house

2. What are the three main types of guest sales transactions?

- ____________________
- ____________________
- ____________________

3. List three of the five types of server productivity reports that can be generated by a POS system.

- ____________________
- ____________________
- ____________________

4. What is a general ledger program?

____

____

5. What are the four basic components of a computerized accounting system?

- ______________________________
- ______________________________
- ______________________________
- ______________________________

6. List four of the six areas of management decision making that computers can facilitate.

- ______________________________
- ______________________________
- ______________________________
- ______________________________

Chapter 14 Exercise

1. What are the four types of business formation?
   - ______________________________
   - ______________________________
   - ______________________________
   - ______________________________

2. What is a feasibility study?

   ____

   ____

3. Why might a quick-service restaurant do better near a public transit center than a full-service restaurant?

   ____

   ____

4. List seven of the fourteen types of questions a good market study should ask.
   - ______________________________
   - ______________________________
   - ______________________________
   - ______________________________
   - ______________________________
   - ______________________________
   - ______________________________

5. What is a net lease?

   ____

   ____

6. What is a net, net lease?

____

____

7. What is a net, net, net lease?

____

____

8. List three types of pre-opening expenses.

- ____
- ____
- ____

9. List three types of working capital.

- ____
- ____
- ____

10. Match each term on the left with its definition on the right.

____(1) Return on Assets	a. Profit left over after expenses and taxes
____(2) Return on Investment	b. Ratio of net profit to net assets
____(3) Net Profit	c. Ratio of direct income to original investment

11. What are the three types of tests that can be used to determine return on investment?

- ____
- ____
- ____

Study Outline

Chapter 1

1. The foodservice industry, a large section of the larger hospitality industry, is a vast and varied field.
2. Evidence has concluded that prehistoric humans prepared and cooked food for large groups, for both social and religious functions. Many ancient peoples—the Assyrians, Chinese, Egyptians, Greeks, and Romans—enjoyed the luxuries of specially-prepared banquets and meals.
3. Advances in European cuisine took place mostly in the courts of royalty and nobility during the Middle Ages. At this time, the professional status of chefs rose.
4. In the early 16th century, during the European Renaissance, Catherine de Medici left Italy to marry King Henry II of France. She brought fine foods, recipes, and chefs to her new home. This served as the beginning of classical French cooking, which continued to develop under future French kings.
5. In the middle 18th century, Boulanger opened a public eatery at which he prepared and sold soups he claimed would restore people's health. He called the soups *restaurers*, which is the origin of the English word, restaurant.
6. From the 17th century on, the French borrowed heavily from other European cuisines, including English, Russian, Italian, and Polish.
7. The 19th century saw one of the greatest massive societal changes in modern history—the Industrial Revolution. As capitalists began to earn money and a wealthier working class emerged, dining out became more popular and fashionable, and restaurants prospered.
8. The "Golden Age of Cuisine" lasted virtually the entire 19th century and several decades into the 20th. Its first "star" was the great Marie Antoine Carême, who became one of the most renowned and accomplished chefs of all time. He perfected the preparation of consomme, and used elaborately carved ices and foods to decorate banquet tables.
9. Later in the 19th century, Georges Auguste Escoffier became the greatest living chef. He worked with the unrivaled César Ritz, who operated the finest hotels in Europe, to serve people in the highest style. He simplified the elaborate series of French classical courses, and wrote many articles and books about cooking.
10. Colonial American eating places consisted mostly of small operations catering to travelers. Beginning around 1850, fine hotels were built in large cities, serving

wealthy travelers. The largest boost to the American foodservice industry came in the 1950s, with the start of "fast food" and chain units. Today these operations dominate the industry in sales.

Chapter 2

1. The foodservice industry is one of the top 10 American—and worldwide—money-makers. In the United States, one-fourth of the food consumed each day is eaten in a foodservice establishment.
2. The National Restaurant Association divides the foodservice industry into three groups. Group I is commercial food service (restaurants, food stands, bars and taverns, foodservice contractors, lodgings, retail hosts, recreation areas, and vending facilities). Group II is institutional food service (employee feeders, schools and universities, transportation, hospitals, and nursing homes). Group III is military food service.
3. In order to earn a profit in commercial operations, foodservice managers are continually challenged to balance cost with income. Costs and sales must always be kept in mind as managers plan and develop menus.
4. There are several types of operations that fall within the commercial group.
   a. Restaurants and lunchrooms account for just over one-third of all commercial operations.
   b. Family-style restaurants are those that offer home-style menu items, a casual atmosphere, moderate prices, and items for guests of all ages.
   c. California-menu operations offer casual breakfast, lunch, dinner, and snack items at all hours of operation.
   d. Quick-service units make up another third of the commercial segment, and account for a huge portion of total sales.
   e. Commercial cafeterias can provide low prices and moderate fare because patrons serve themselves.
   f. Social caterers prepare food in one location and serve it in another, usually for special events.
   g. Frozen dessert units have relatively simple menus, low costs, minimal service, and must, therefore, depend on high volume.
   h. Bars and taverns are much more likely to be independently owned than restaurants. They are more likely today to serve food and provide entertainment than their counterparts several decades ago.
   i. Food contractors provide food services to businesses, transportation companies, and institutions that cannot provide their own.

- j. Large lodging places often provide several food services, from fine dining, to snack bars, to room service. Smaller ones may provide only a coffee shop.
- k. Retail hosts are establishments such as grocery stores, highway rest stops, and convenience stores. Low prices and individual selection of items have made these a larger competition to other foodservice operations in recent years.
- l. Recreational facilities provide food and beverages to people at sports stadiums, amusement parks, concert facilities, racetracks, fairs, and a host of other entertainment sites.
- m. Mobile caterers sell packaged items to employees, construction workers, and passers-by.
- n. Vending services can be found anywhere people want low-priced fast foods.

5. There are several types of operations that fall within the institutional group.
   - a. Employee feeding comprises those contract companies that serve employees in plants and office buildings. They usually include cafeterias, vending operations, and sometimes executive dining rooms.
   - b. Public elementary and secondary schools must provide student lunches that follow nutritional guidelines set by the U.S. Department of Agriculture. Private schools with boarders must provide three meals daily to students.
   - c. Colleges and universities often provide meals in cafeterias as well as foods in student unions and meals for special occasions. College food services must cater to students' unique tastes and schedules.
6. The transportation feeding group includes several types of operations.
   - a. Airlines must provide meals to passengers under especially challenging circumstances. Service must be limited to a short amount of time, so food must be preprepared, and sanitary considerations are critical. Airlines often use cycle menus.
   - b. Railroad and bus lines normally provide simple foods and vending services, both on trains and buses and in stations and waiting areas.
7. Food service in health-related institutions accounts for just over five percent of total foodservice sales.
   - a. Hospitals alone bring in more than $10 billion each year. Their foodservice operations usually are huge and rigidly complex. They generally use some type of cycle menu, and must meet the many dietary needs and requirements of patients while still satisfying the tastes of their "customers." Sanitation and food safety are critical. Many hospitals use centralized services in which foods are prepared in one location and transported throughout the facility to patients. Efficient cooling and heating units are crucial.

   b. Long-term healthcare facilities are those that provide health care in a home-like environment. This includes nursing homes and hospices. Nutrition, special dietary needs, and sanitation are crucial, as with hospitals.

8. Clubs exist for a great variety of purposes: social groups, athletics, university alumni groups, and veterans' groups. Most clubs emphasize catering to all the needs of their guests. This usually includes several food services, from a bar, lunch room, and dining room, to poolside and workout-room facilities.
9. The military is responsible for feeding hundreds of thousands of people every day. Besides daily meals for enlisted soldiers, the military operates officers' clubs and facilities for soldiers' families.
10. Central commissaries are large areas of product receiving and/or production from which foods are shipped out to individual units. Commissaries are used by many quick-service chains and large institutions such as universities. The degree of centralization depends on the facility.
11. Many external forces have and will continue to have an enormous impact on the foodservice industry.
    a. Economic hardships and unpredictability make it necessary for companies to be ever-more competitive in their efforts to attract and retain customers.
    b. Changes in birth rates, the elderly population, and family structures will draw a different picture of "typical" foodservice customers, their spending, and their dining patterns.
    c. The American work force will increasingly be made up of recent immigrants and women. Their needs, expectations, and attitudes are certain to be different from their white male counterparts of 20 and 30 years ago.
    d. Americans have become more and more conscious of their nutritional needs and they are willing to pay for nutritious menu choices. While it is true that many customers "cheat" when they eat away from home, they patronize those operations that allow them to choose nutritious choices.
    e. As the makeup of the work force changes, problems with employee turnover—which is especially high in the foodservice industry—will continue. Profitable operators and managers will need to implement impeccable recruiting, hiring, motivational, and developmental techniques in order to attract and keep the best employees.
    f. Employers will also have to meet the challenge of managing a culturally diverse group of employees. Differences in age, ethnic background, educational level, religion, gender, and sexual orientation will make for rich organizational cultures that managers will be responsible for overseeing. In addition, more people than ever before have become aware of their legal

rights at work, and managers and employers are urged to actively prevent legal problems rather than merely reacting to them later.

g. Government regulation of sanitary practices, the minimum wage, safety, taxes and employee benefits, civil rights, smoking, alcohol-related accidents, and menu claims will increase, and managers must stay abreast of all federal, state, and local changes to the law.

12. Internal factors within the foodservice industry will also have an impact on its future developments. Menu choices, competition for patrons, and promotions have become increasingly creative within the foodservice industry.
13. Leaps in technology promise to help operators become more efficient. These include energy-efficient cooking equipment, improvements in packaging, and changes in food production such as genetic engineering and controlled farming.

Chapter 3

1. Menus, in any type of operation, must serve two functions: as a planning and control document for the back of the house, and as a means of communication to patrons.
2. A well-planned and designed menu guides patrons into making the best choices for their meals, and communicates the mission, theme, and history of the operation.
3. Menu planners must have a clear understanding of the long-term (strategic) and short-term (tactical or operational) goals of the organization. The most basic goals are written in a mission statement, or concise description of the organization.
4. Operators must set budgets, managers must adhere to budgets, and both must analyze and compare budgeted figures with actual figures to determine the operation's efficiency. Budgets are plans that tell managers how well the business is doing.
5. One of the first steps involved in menu planning is the number of menus an operation will require. This depends on the size and scope of the operation, its mission, and its clientele.
6. Next, menu planners must determine the type(s) of menu to be used. The following are some common menu types.
   a. A la carte menus offer items separately at a separate price.
   b. Table d'hôte menus group several items together at a single price.
   c. Du jour menus feature items served on a particular day.
   d. Limited menus offer only a few specialized items.

e. Cycle menus repeat items offered after a set amount of time.
f. California menus offer breakfast, lunch, dinner, dessert, and snack items at any time during the hours of operation.

7. Menu planners should keep in mind the common sequence of menu items when choosing what to offer. Breakfasts are traditionally lighter than lunch and dinner, although a few substantial items should be included on a breakfast menu. Lunch is generally lighter than dinner, and traditionally includes sandwiches, salads, soups, and lighter entrees. Americans, unlike many other cultures, eat their heaviest meal in the evening, and most menus reflect this. The sequence of foods will depend on the type of operation and its clientele.
8. Item variety is essential to most menus. Contrasts—spicy and mild, hot and cold, crunchy and smooth, hearty and light—should be available.
9. The number of items chosen for a menu varies from operation to operation, and any number should be based on patrons' preferences.
10. The occasion of a menu helps dictate its item selections.
    a. Breakfast menus can contain any number of unique items, but there should be at least a few that customers expect, such as eggs, omelets, breakfast meats, pancakes or waffles, and toast and pastries.
    b. Brunch menus combine both breakfast and lunch items and are intended for late morning and early afternoon. Lunch menus should offer both light and hearty selections.
    c. Afternoon menus can include light lunch items, snacks, and desserts.
    d. Dinner menus are traditionally the most elaborate in the United States. Entrees usually include a main dish (either meat-oriented or meatless), a starch, a vegetable, and a salad. Beverages precede, accompany, and follow dinners. Desserts may or may not follow. Many formal dinner menus follow the classic French sequence of courses.
    e. Evening menus must appeal to people who are out late. They may include snack items, desserts, and fancy bar drinks.
    f. Menus for special occasions—birthdays, holidays, social gatherings, weddings, and parties—must be unique and memorable, and must, above all, please the client.
    g. Menus for high tea include light desserts and small sandwiches. The British enjoy both high and low teas as a light meal between lunch and a late dinner.
    h. Buffet menus are meant to impress the eye as well as the stomach. Variety is essential, as is using foods that will hold up well for the duration of the event.

11. Many operations find it profitable to provide menus for special customer groups.
    a. Children's menus should offer a manageable number of small portions of foods that appeal to young palettes. The shape and look of the menu should also appeal to children.
    b. Menus for older adults should reflect variety and common dietary limitations, such as lowered salt, lowered fat, softness of texture and consistency, and moderate flavoring.
12. Menus for institutional or noncommercial operations must appeal to customers in many of the same ways commercial menus must, even though profit is not the main financial motive. Variety is especially important in facilities in which patrons are obliged to take all their meals for an extended amount of time. Special events and themes can help break monotony. Nutritional considerations are important not only in healthcare facilities, but in all facilities in which patrons have little choice of where they eat.

Chapter 4

1. Menu planners cannot create and design menus in a vacuum. They are limited somewhat by factors dictating their future success.
2. Physical control factors include the following.
    a. The operation's equipment
    b. The layout of the facility
    c. Labor and time available to prepare and serve foods
    d. Financial constraints and goals
    e. Product availability, quality, and cost
    f. Events and seasonal changes in customer volume
3. Market control factors include the following.
    a. Nutritional needs and demands of patrons
    b. Appearance, temperature, texture, consistency, and flavor of menu items
    c. Variety, complexity, and perception of taste
    d. Guest expectation and perceptions
4. It is extremely important to know patrons' food preferences before developing the menu. Many operators use taste panels made up of potential customers or employees, or both, to test possible menu items for their flavor, consistency, texture, appearance, temperature, and general appeal. Panel members are asked to rate different items in order of their preference. Operators can use this information to help them make menu decisions. The people used on a taste panel

should be representative of the operation's clientele, and conditions—weather, individuals' health, time of day, time allotted, etc.—should be favorable.

Chapter 5

1. An operation cannot operate successfully—and a menu will not perform effectively—without carefully planned and executed cost-control measures. These must be followed during every step of operation: purchasing, receiving, storage, issuing, preparation, and production.
2. One of the most basic foodservice calculations is total cost of goods used or total food cost. It is calculated as follows.

   Beginning (Opening) Inventory
   \+ Purchases
   – Ending (Closing) Inventory
   = Cost of Goods Used (Food Cost)
3. A daily food cost report helps an operation keep track of food used each day. Actual daily food cost and overall cost of goods sold are often predicted and compared with calculations. This is a function of operational budgeting.
4. To find food cost for menu items, whole portions or recipes are estimated and then divided by their number of portions, or yield. A yield test gives the ratio of usable food to trim, shrinkage, and unusable parts. With this information, managers can determine how much as purchased (AP) amount of an item is needed to yield an edible portion (EP) amount.
5. Labor costs must also be considered and controlled. Labor cost will be high if many items are prepared from scratch, and relatively low if many preprepared convenience items are used. Training and positive work conditions can raise productivity and lower labor costs.
6. Direct labor is the labor cost related to handling, producing, and serving a menu item. The direct labor cost plus an item's food cost gives the item's prime cost.
7. Both commercial and noncommercial operations use a food cost percentage to keep food costs down. The food cost percentage is food cost divided by sales. Common food cost percentages are between 25 and 40 percent. A profit-and-loss (income) statement shows food costs along with other costs and revenue for a set period of time.
8. Food and labor costs can be largely controlled by accurate forecasting of business volume and production needs. Managers should look at sales histories, weather

predictions, planned promotions, and related events when predicting sales volume.

9. Portion control keeps costs down by minimizing waste. Equipment and measuring tools should be used, as well as standardized recipes for all menu items. Training helps employees stick to standard portions.
10. Labor costs are controlled through the use of employee schedules and labor costs per menu item. Production work schedules should make clear the following points.
    a. The period of time covered
    b. The tasks to be done
    c. Who is to perform the tasks
    d. How much of each item is to be produced
    e. Which recipe numbers are to be used
    f. Portion sizes
    g. Time allotted for task completion
    h. Tasks to be completed during slow periods
11. Work schedules should always be compared to actual hours worked to find discrepancies and avoid overpayment of employees.
12. Managers can raise employees' productivity by using effective methods of recruiting, hiring, and training that ensure the most qualified people are chosen and assigned to the most appropriate tasks. Tasks can be completed more quickly and easily by minimizing unnecessary movement, which can enhance the job itself.

Chapter 6

1. The more well-planned and practically-based menu prices are, the more likely they are to meet customers' expectations and bring in profits.
2. It is crucial that customers feel they are receiving suitable quality and value for their money. If low prices are not an operation's primary attraction, the operation must convince customers that its prices are warranted. Most consumers will pay for well-differentiated products.
3. A selling price of $6.95 or $6.99 is perceived as substantially lower than $7.00. Prices that end in 5 are the most common. Prices from $0.86 to $1.39 are considered to be about $1.00; from $1.80 to $2.49, about $2.00; from $2.50 to $3.99, about $3.00; from $4.00 to $7.95, about $5.00. Menu planners are wise to raise menu prices within the same range rather than into a new, higher range.

4. All menu prices should be based on market research. This research should take into account customer preferences and perceptions, seasonality of foods, competitors' prices, special promotions, and external economic conditions.
5. One pricing method is to add a desired food cost percentage to an item's food cost. For instance, if an item's cost is \$3.78 and a 36 food cost percentage is desired, selling price would be \$3.78 ÷ 0.36 = \$10.50. The same selling price is calculated using a pricing multiplier or pricing factor: \$3.78 ÷ (1.00 ÷ 0.36), or 3.78 × 2.78 = \$10.51. The drawback of these methods is that they only account for food cost.
6. A menu item's prime cost is its food cost plus cost of direct labor needed to produce the item. The prime cost can be divided by the desired cost percentage to arrive at a selling price.
7. Some operations use an all or actual cost pricing method in which food cost, labor cost, operating cost, and desired profit (0 for a noncommercial or a break-even operation) are added for a selling price. This is similar to the cost-plus-profit method.
8. In gross profit pricing, the gross profit figure is divided by the number of guests to compute an average profit per guest.

Chapter 7

1. Since the principle functions of a menu are to communicate and sell items, the look, feel, and design of a menu are crucial. If a paper menu is used with a sturdy cover, the operation's name and logo should appear on the cover.
2. Production menus are designed for production employees rather than customers. They list menu items, ingredients, recipe references, amounts to prepare, and portion sizes.
3. The menu should be written using a typeface that is easy to read and fits the aims of the operation. Serif typefaces have distinct edges; sans serif typefaces do not. Regular type is easiest to read, but bold, italic, and script type can be used for special effects and to emphasize certain items.
4. Type is measured in points. 10-point and 12-point type are common for normal reading, while 18-point type is common for headings. Spacing of letters, spacing between lines of type (called leading), and weight of type add to the ease of reading.
5. Writing on menus should be neither bunched together nor too spread out. Margins require considerable space.

6. Where items are placed on the menu can influence how often they are chosen by customers. The first and last items in a column are seen first and most prominently; items in the middle are noticed less. Mixing up items by price forces customers to look at all the items. Menu items that the operation wishes to sell should be placed in emphasis areas, which are at the right of a two-page menu and in the middle of a one-page menu.
7. Color can be used to achieve almost any effect on a menu. Colors should foster easy reading, and reflect the theme and mood of the operation.
8. The menu cover should be durable, made of heavy paper, and covered with some type of lamination that can be kept clean and protect the interior. Interior pages can be made of paper with either a smooth or rough texture.
9. The shape of a menu can also add to its appeal and ability to sell items.
10. Whether menus are printed professionally or in-house using a desktop publishing system, printing large amounts can save the operator money. Unpriced menus can be kept at the printer's so price changes can be made when new menus are needed.

Chapter 8

1. No menu will be successful unless managers continually analyze how well it is doing its job and how well items are selling. There are seven common ways to analyze menus.
   a. Menu counts can be kept manually, but most counts today are kept by computerized point-of-sale machines.
   b. A subjective evaluation allows managers to gather feedback based on menu experts' judgments.
   c. A popularity index indicates the popularity of menu items in terms of number sold, sales, profit margin, and how they cover their costs. Each item receives an expected index number based on how many members are in the group of items being compared. If five items are being compared, the expected index is 0.20 ($1.00 \div 5.00 = 0.20$). Since it is difficult to compare the index numbers of items in groups with different numbers, items are given a popularity factor by dividing the actual index by the expected index.
   d. A menu factor analysis compares menu items' popularity, sales contribution, food cost, and gross profit contribution. (See text pages 195–198 for examples.)

e. The Hurst method of analyzing menus consists of 18 steps leading to a menu score that can be used for comparison. The higher the menu score, the greater potential for profit. (See text pages 198–202 for examples.)
f. Goal value analysis sets up a series of calculations based on food cost percentage, number sold, selling price, and variable costs. Actual values are compared to goal values. (See text pages 202–203 for examples.)
g. A menu that breaks even covers all costs associated with it. Break-even analysis is used primarily by noncommercial operations and new commercial operations that will not yet earn a profit. Calculations will tell operators how many dollars in sales are needed to break even, and how many customers are needed. Similar calculations are made to show the level of business volume in which a certain profit will be earned. The calculation is: break-even point = fixed costs divided by (average check times fixed cost percentage). (See text pages 203–205 for examples.)

2. Productivity reports and daily food cost reports aid in analyzing menus.

Chapter 9

1. Increased awareness of the health and safety dangers associated with drinking has led to an overall decrease in alcohol consumption among Americans. As a result, food and drink operators have had to promote responsible alcohol service, as well as low-alcohol and nonalcoholic drinks on their menus.
2. In order to reap profits, the liquor menu must be promoted by servers, all employees must follow established procedures, and items must match patron desires and expectations.
3. Alcohol sales regulations are increasing, as well as the liability of operators and servers for alcohol-related injuries. Every operator must know and adhere to state laws.
4. It is not necessary for liquor menus to be elaborate in order to meet needs. A short, simple, but well planned menu can bring in healthy profits.
5. Menus can be divided according to liquor classifications: cocktails, mixed drinks, beers and ales, wines, and after-dinner drinks. (See Exhibit 9.1 in the text for examples of each.)
6. Beers, wines, and spirits should match the theme, food, decor, and clientele of the operation. For example, a Mexican restaurant would most likely offer Margaritas, Mexican beer, and Spanish wine.

7. Many wine lists include a house wine that the operation buys in bulk and serves in carafes or by the glass; it is offered along with bottled wines. All wines should be named and described accurately.
8. Traditionally, red dinner wines are served at room temperature (60°F to 65°F, or 16°C to 18°C); blush and white dinner wines are served chilled (about 45°F or 7°C). However, many operators give priority to patrons' individual preferences and tastes, even if they go against conventional wisdom and tradition. Servers should be able to suggest suitable wines for different entrees.
9. Information and interesting anecdotes about wines can be effective promotional tools.
10. Many spirits lists contain only those items featured by management, since most patrons assume they can receive almost anything they would desire from the bar.
11. American proof is twice the alcoholic percentage of a liquor. If a spirit is 50 percent alcohol, it is 100 proof.
12. Operations should offer a variety of beers and ales, including some light beers and a local or regional favorite.
13. A sommelier, or wine steward, is on staff in some fine-dining establishments to answer questions, suggest wines, and take the order. If there is no sommelier, servers must fulfill this role.
14. Proper wine service is very important. Wines should be served in appropriate glasses. A small amount is poured into the hosts' (the person who ordered the wine) glass to taste. If it is acceptable, the wine is served to all of the other guests, then to the host.
15. Wines are normally given a straight markup, usually between 50 percent and 100 percent. Wine is priced based on competitors' prices and what the market will bear, much more than food is. Mixed drinks are often priced according to portion costs taken from bottle costs. (For instance, if a 32-ounce bottle of scotch costs $16.89, each ounce has a portion cost of $0.53.) Tap beers are also priced according to portion costs and pouring loss.
16. Liquor and other beverage costs must be controlled. Appropriate quantities and qualities are purchased through purchase orders. Both understocking and overstocking can cost an operation money. All deliveries should be checked carefully, and stock items should be issued only to employees with a signed requisition form.
17. Much waste and spillage can be controlled through the use of standard bar recipes. Measurement utensils, such as jiggers and shot glasses, also reduce waste.

18. A formal system of guest checks should be used to discourage mistakes and dishonesty by employees and guests. Numbered checks are issued to servers, who are responsible for all payments recorded on them. Most electronic cash registers keep an automatic count of drinks sold and money taken in through sales.
19. The standard deviation and drink differential methods of predicting sales from liquor are used to control purchases and sales. (See text pages 235–239 for examples.)

Chapter 10

1. More and more Americans say they are concerned with healthy lifestyles and nutritional eating. The foodservice industry has capitalized on this development by offering healthful menus to meet many different patron desires and needs.
2. The federal government has passed recent regulations for labeling foods and menu items. Foodservice operators can promote their concern for health issues by featuring healthy menu items labeled according to government regulation.
3. Operators can offer more nutritious menu items by altering existing recipes and by developing new items. Some ways to incorporate more nutritious items include offering ethnic foods that are low in fat, natural or organic, and appeal to vegetarians.
4. The USDA has developed an Eating Right pyramid to replace its conventional four food groups. The pyramid shows proportions, from largest to smallest, of the diet that should come from six groups of foods: breads, fruits, vegetables, proteins, milk products, and sweets. (See Exhibit 10.2 in the text.)
5. Menu planners should have a general knowledge of the basic nutrients, their importance to the body, and their sources in the diet. The basic nutrients are grouped as carbohydrates, proteins, fats, vitamins, minerals, and water.
6. From 55 to 65 percent of a person's total calories should come from carbohydrates, which provide the body with energy in the form of simple and complex sugars. Dietary sources include potatoes, pasta, rice, legumes, and sugary foods.
7. Protein is found in animal foods—meat, fish, milk, eggs, cheese—and nonanimal foods, especially legumes, nuts, and grains. Proteins also can be made by combining foods, such as legumes and grains (as in red beans and rice). Protein is needed for the body's growth and muscle formation.
8. Fats are needed for normal body functions and to protect internal organs. Only 30 percent of our calories should come from fat; most Americans eat too much fat.

Polyunsaturated and unsaturated vegetable oils are generally more nutritional than saturated fats found in animal foods. Eating fatty foods can lead to increased amounts of cholesterol in the blood and buildup of plaque along artery walls. This can lead to a heart attack or stroke.

9. The body uses only small amounts of each vitamin to perform essential functions.
   a. Vitamin A helps healthy functioning of the eyes. Lack of it can lead to vision problems. It is found in leafy green vegetables, yellow and orange fruits and vegetables, sweet potatoes, and cheese.
   b. Vitamin B_1 (thiamin) promotes the appetite and burns energy. Insufficient supply of this vitamin can lead to the disease called beriberi. It is found in whole grains, pork, liver, legumes, and milk.
   c. Vitamin B_2 (riboflavin) supplies energy, and without it, sores can develop. It is found in milk, cheese, eggs, meats, and whole grains.
   d. Niacin (a B-complex vitamin) supplies energy, and lack of it causes the disease called pellagra. It is found in peanuts, seeds, beer, meats, and whole grains.
   e. Vitamin B_{12} helps prevent anemia and promotes healthy blood. It is found in most foods, especially fish, poultry, meats, cereals, dairy products, and nuts.
   f. Folic acid, or folacin (a B-complex vitamin) helps form blood. It is found in liver, legumes, asparagus, and broccoli.
   g. Vitamin C helps form healthy tissue, bones, blood, and teeth. Vitamin C deficiency leads to scurvy. It is found in citrus fruits, tomatoes, cabbage, berries, and many vegetables.
   h. Vitamin D helps form healthy bones and teeth, and a lack of it causes rickets. It is made by the skin when it is exposed to sunlight, and it is found in fish liver oils, cream, butter, and eggs.
   i. Vitamin E helps muscles and other processes function. It is found in whole grains and vegetable oils.
   j. Vitamin K helps the blood coagulate and is found in most fresh fruits and vegetables.

10. Minerals help regulate many of the body's vital processes.
    a. Calcium is important in bone formation, and is found in milk and other dairy products.
    b. Phosphorus is found in meats and other protein-rich foods.
    c. Iron is important for healthy blood, and a lack of it can cause anemia. It is found in green leafy vegetables, meats, egg yolks, prunes, whole grains, and legumes.

d. Copper also is needed for healthy blood, and is found in many commonly eaten foods.
e. Sodium is most commonly found in table salt. It is needed for proper muscle formation, but an excess can exacerbate high blood pressure.
f. Potassium is found in bananas, dried apricots, and orange juice, and can be lost through perspiration. It helps the body maintain its water and acid-base balance.
g. Trace minerals, including magnesium, sulfur, and zinc are needed only in very slight amounts.

11. Water is necessary to keep many bodily functions operating properly and to flush out toxic substances. Adults should drink six to eight glasses a day.
12. Although it is not considered a nutrient and it is not digestible by humans, fiber is necessary for proper digestion and clean functioning of the intestines. It is found in fruits and vegetables.
13. Cook small amounts at a time, using cooking methods, such as steaming, baking, and broiling. This will preserve more nutrients in food during cooking and preparation. Foods should be stored only for short times before serving. Vegetables should be held no longer than 20 or 30 minutes in a steam table. (See the five steps for preparing nutritional foods on text page 260.)
14. Many Americans have adopted low-calorie, low-fat, low-cholesterol, low-sodium, and low-sugar diets. Promote items fulfilling these dietary requirements on the menu.

Chapter 11

1. A well-developed menu dictates the items, quantities, quality, and amounts that should be purchased by an operation. The steps involved in foodservice purchasing are as follows.
   a. Determine the need for an item and what quality, quantity, etc. are required.
   b. Search for the item on the market.
   c. Negotiate with suppliers for the best price.
   d. Receive the correct items.
   e. Evaluate the purchasing process for efficiency.
2. Since markets change frequently and an operation's needs may change, the purchaser must be extremely knowledgeable about the buying environment.
3. Amounts needed are determined by sales histories and forecasts, and by reasonable reorder points (ROPs).

4. A formal buying process involves the purchaser collecting written bids from potential suppliers. Informal buying involves verbal quotes and negotiations.
5. Specifications can help delineate exactly the type of item needed by an operation. The specification for an item lists amount needed, brand name, packaging size, quality needed, price desired, grade, and other factors.
6. Receiving, storing, and issuing are all considered phases of the purchasing function, and at all of these stages control must be strictly maintained.
   a. All deliveries should be verified against the original purchase order and the supplier's invoice.
   b. Storage temperatures must ensure little or no waste. Security should be maintained.
   c. The first-in, first-out (FIFO) method of storage rotation ensures that older items are used before newer ones.
   d. Items are issued to employees with careful control and reordered before they run out.
   e. Careful inventories—either perpetual, physical, or both—are taken to control stock items.
7. The most carefully arranged menu can be undermined if items are prepared carelessly or inconsistently.
8. Daily forecasts of item amounts, called production sheets, are based on sales histories and forecasts.
9. Standard recipes help ensure product quality and proper quantity, while keeping product costs down. Employee training is key.

Chapter 12

1. Only well-trained employees with positive attitudes can deliver all that the menu promises.
2. All customers, in any type of operation, expect to be greeted warmly, promptly, and sincerely. In seated operations, they expect to be seated within a reasonable amount of time.
3. Servers must not only be friendly, but efficient in the way they take orders, convey orders to the kitchen, deliver items in a timely manner, meet customer needs throughout the meal, present the check, and receive payment. All guests should be thanked, bid good-bye, and invited to return.

4. Operations are largely defined by the type of service they offer.
   a. Counter service is fast and encourages high customer turnover. It does not necessarily save space, and requires an efficient work setup.
   b. In cafeteria service, customers serve themselves and carry a tray to a table. Customers may bus their own tables or have them bussed by employees. Space is saved if customers go to different areas for different foods.
   c. Buffet service is used for special meals and special occasions. Customers move along the table and are served by employees. Beverages are usually served at tables. Much food can be wasted at buffets, but food must be appealing both to taste and sight.
   d. Seated service can lend itself to relaxed family-style dining or very formal fine-dining.
5. Within the category of seated service, American, French, Russian, and English service each conjure different moods and themes.
   a. In American service, food is dished onto plates in the kitchen and carried out to guests by servers. It is generally less formal than French or Russian service.
   b. French service is characterized by a legion of servers, led by a chef du rang, or captain, who serve all foods from a gueridon, or tableside cart. Much skill and finesse is required of servers in French service.
   c. Russian service is very elegant and demonstrates servers' flare. Food is brought out on platters and in large bowls and served to guests by servers. Each course gets a completely new setting.
   d. English service is a type of family service in which a host seated at the guest table serves guests from platters and large bowls, often carving a large roast or bird of some kind. This type of service is relatively uncommon but might be used in a family-run inn.
6. All guest checks should be totalled accurately and presented only after guests have completely finished their meal. If it is not known who will be paying, the server should lay the check in the middle of the table.

Chapter 13

1. Computers have important applications in the front of the house, back of the house, and in managerial planning and decision making.
2. In the front of the house, computerized point-of-sale systems have made guest checks more efficient and accurate. Inefficiency and dishonesty of employees is much less likely with a computerized system. Orders are transmitted to the

kitchen quickly and accurately, and sales are easily reconciled at the end of the day and can be used for more long-term analysis later.

3. Computers can produce server productivity reports so that servers and managers know how well each server is performing.
4. Daily sales information entered into the computer can be analyzed for longer periods of time. These actual figures can be compared to budgeted figures and used to predict future sales.
5. Computers also make tasks more efficient in the back of the house. Purchasing is aided by computer systems that keep track of past purchases, reorder points, inventories, and specifications.
6. Computerized accounting procedures are common in the foodservice industry. Budgets and financial reports, such as income statements and balance sheets, are easily compiled from computerized information.
7. Management planning and decision making are made effective through the use of computerized data and reports. Long-term investments and financial decisions can be based on solid figures, and menu analysis by computer offers much valuable information.
8. As with any computer application, a computer in a foodservice operation will perform only as accurately as the information that is put into it.

Chapter 14

1. Investment in food services is thought by lenders to be a risky venture because the rate of failure is high and costs can easily outweigh sales and profits. Many new operations require an initial investment of $500,000 to $1 million.
2. Before lenders will invest in a new operation, they request a detailed feasibility study, which outlines such things as the overall concept of the operation, its location, the target market, the local community, the initial operating budget, pro forma (forecasted) income statement and balance sheet, and other information.
3. Once investment is obtained, many financial decisions must be made by owners and managers, such as whether to own, rent, or lease the facility and its equipment.
4. Financial analysis is ongoing, once an operation is up and running. The return on investment (ROI) tells managers their return on capital investments, expressed as a percentage of the original investment. For instance, if an operator invests $200,000 in a new business and earns a profit of $20,000, the ROI is 10 percent.

5. Cash flow figures tell managers and investors how much money is available for operations, as opposed to money tied up in inventory, investments, etc. Cash flow is essential so that bills are paid on time, employees are paid, etc.
6. Liquidity refers to a business having convertible (liquid) assets, or those that can be converted rather quickly into cash, that exceed its liabilities, or debts. Several ratios help managers assess liquidity.
   a. The current ratio compares up-to-date assets to current liabilities. A ratio greater than 1:1 (2:1, 3:1, etc.) is desirable.
   b. Another ratio compares sales in dollars to working capital (current assets less current liabilities). High ratios ($40:1, $50:1, etc.) are desirable.
   c. The solvency ratio compares owners' and stockholders' equity to money that is owed to creditors. A ratio of 1:1 is acceptable, but higher ratios (2:1, 3:1, etc.) are preferred, and low ratios (1:2, 1:3, etc.) indicate large debts.

Practice Test

This Practice Test contains 80 multiple-choice questions that are similar in content and format to those found on The Educational Foundation's final examination for this course. Mark the best answer to each question by circling the appropriate letter. Answers to the Practice Test are on page 75 of this Student Workbook.

Lesson 1: Overview of the Foodservice Industry

1. The oldest existing cookbook was written by Apicius in ancient
   A. Greece.
   B. Egypt.
   C. Rome.
   D. Assyria.

2. A toque is a chef's
   A. knife.
   B. hat.
   C. Blue ribbon.
   D. guild.

3. Georges Auguste Escoffier was the most accomplished chef of which century?
   A. 17th
   B. 18th
   C. 19th
   D. 20th

4. The first hamburger chain, appearing in the 1930s, was
   A. McDonald's.
   B. Burger King.
   C. White Castle.
   D. Wendy's.

5. Every day, what percentage of the food consumed by Americans is served by employees of foodservice operations?
   A. 15 percent
   B. 25 percent
   C. 35 percent
   D. 45 percent

6. In which type of operation are beverage alcohol sales likely to be the most significant?
   A. Quick-service
   B. Employee contract cafeteria
   C. Full-service restaurant
   D. Nursing-home cafeteria

7. Which of the following operations is considered part of the commercial segment of the industry?
   A. Sports stadium
   B. High school
   C. Business and industry contractor
   D. Military

8. Which type of menu offers breakfast, lunch, and dinner items during all hours of operation?
   A. California
   B. Limited
   C. A la carte
   D. Du jour

9. Which of the following types of operation is most likely to employ an executive chef, a sous chef, and a chef du parti?
   A. Family-style restaurant
   B. California-menu restaurant
   C. Private club
   D. Hotel food service

10. Which of the following would be considered part of the retail host segment of the foodservice industry?
    A. Frozen dessert unit
    B. Quick-service operation
    C. Convenience store
    D. Lodging food service

11. The majority of mobile caterers sells food and beverages to
    A. outdoor and office workers.
    B. spectators at sporting events.
    C. people in parks for recreation.
    D. patients in hospitals and nursing homes.

12. A central commissary is a place where
    A. employees in a corporation's headquarters are fed.
    B. food is mass-produced before being shipped to individual sites.
    C. food and beverages are sold to individual chain units.
    D. military officers and noncommissioned officers are fed and entertained.

13. In the future, a majority (65 percent) of food service workers are expected to be
    A. women.
    B. teenagers.
    C. people over age 55.
    D. White.

Lesson 2: Developing the Menu and Cost Controls

14. Which of the following is the best example of a strategic plan?
    A. Purchase order for next week's inventory
    B. Daily production schedule
    C. List of specials for the week
    D. Promotional budget for the next two years

15. "We will serve nutritional breakfasts and lunches in accordance with federal guidelines." This is most likely the mission statement of which of the following foodservice operations?
    A. Quick-service operation
    B. Family-style restaurant
    C. Business-and-industry contract food service
    D. Elementary school

16. Which type of menu offers items separately at separate prices?
    A. California
    B. A la carte
    C. Table d'hôte
    D. Cycle

17. Which of the following operations is most likely to use a limited menu?
    A. Fine-dining restaurant
    B. Quick-service operation
    C. Hospital
    D. Elementary or secondary school

18. Cycle menus are most often used in
    A. quick-service operations.
    B. fine-dining restaurants.
    C. institutional food services.
    D. sports and entertainment facilities.

19. Breakfast menus are generally characterized by which of the following?
    A. High volumes and high profits
    B. Low volumes and low profits
    C. Low volumes and high profits
    D. High volumes and low profits

20. Which of the following colors is thought to enhance the appearance of food?
    A. Purple
    B. Olive green
    C. Pink
    D. Mustard-yellow

21. People's sense of taste is sharpest between the ages of
    A. 10 and 15.
    B. 20 and 25.
    C. 40 and 45.
    D. 55 and 60.

22. Which of the following offers the best taste contrast to spicy, crispy fried chicken?
    A. Chips and salsa
    B. Stuffed jalapeño peppers
    C. Barbecued potato chips
    D. Mashed potatoes

23. A foodservice operation opened the month with $8,716 in inventory, closed with $9,188, and had invoices for purchases totaling $11,276. What was the operation's cost of goods used for the month?
    A. $472
    B. $2,560
    C. $10,804
    D. $11,276

24. The operation in question 23 had monthly sales totaling $29,440. What was the food cost percentage for the month?
    A. 34.8 percent
    B. 36.7 percent
    C. 37.9 percent
    D. 40.1 percent

25. A recipe for Chicken Cordon Bleu costs $18.72. Its portion cost is $1.56. How many portions does the recipe yield?
    A. 10
    B. 12
    C. 14
    D. 16

26. A yield test of a cut of meat tells managers
    A. the ratio of edible meat to fat, bone, and trim.
    B. the cut's portion cost.
    C. how long the meat has been aged.
    D. how old the animal was at the time of slaughter.

Lesson 3: Menu Pricing, Mechanics, and Analysis

27. Setting a product or service apart from competitors' is known as
    A. a markup.
    B. differentiation.
    C. value perception.
    D. breaking even.

28. Which of the following is the most customary selling price to include on a full-service menu?
    A. $6.00
    B. $6.25
    C. $6.80
    D. $6.95

29. A restaurant recently lowered the prices of its least-ordered menu items. However, patrons still have not ordered these items. This restaurant's market is most likely
    A. inflexible.
    B. flexible.
    C. steady.
    D. fluctuating.

30. A food service's à la carte menu item has a food cost of $4.34, and the operation's desired food cost percentage is 34 percent. Based on this, what should the item's selling price be?
    A. $10.95
    B. $11.95
    C. $12.95
    D. $13.95

31. All-you-can-eat promotions are best used with foods that are
    A. expensive.
    B. plentiful.
    C. scarce.
    D. hot or spicy.

32. An operation's total sales for a period were $453,697.68 and the total food cost was $73,358.19. What was the operation's gross profit?
    A. $367,884.09
    B. $380,339.49
    C. $399.765.11
    D. $431,076.44

33. If the operation in question 32 served 83,245 guests, what was the average gross profit per guest?
    A. $4.57
    B. $6.78
    C. $8.99
    D. $10.55

34. A cover charge is a form of which of the following?
    A. Maximum charge
    B. Minimum charge
    C. Average charge
    D. Cost-plus-profit charge

35. In the pricing method based on sales potential, which of the following combinations of menu item factors would translate to three pluses (+ + +)?
    A. High cost, high risk, low volume
    B. High cost, low risk, low volume
    C. Low cost, low risk, high volume
    D. Low cost, high risk, high volume

36. It is most important that an operation's menu reflect which of the following?
    A. Operating costs
    B. Theme and atmosphere of the operation
    C. Number of servers on duty during each shift
    D. Competitors' business philosophies and how they differ from yours

37. Which of the following is considered a disadvantage of script and italic typeface?
    A. It conveys too casual a tone and atmosphere.
    B. It is more expensive to print than plain type.
    C. It is more difficult to read than plain type.
    D. It tends not to emphasize content to readers.

38. A menu item is one of eight competing for a popularity factor analysis. What is the item's expected popularity index?
    A. 0.125
    B. 0.250
    C. 0.300
    D. 0.325

39. If the item in question 38 has an actual popularity index of 0.175, what is the item's popularity factor?
    A. 0.20
    B. 0.75
    C. 1.40
    D. 2.00

Lesson 4: The Liquor Menu and Planning a Healthy Menu

40. Regulation of alcohol sales and dram shop laws fall under the jurisdiction of which of the following lawmakers?
    A. Federal
    B. State
    C. City
    D. County

41. Which of the following is considered a cocktail?
    A. Vodka and tonic
    B. Whiskey sour
    C. Scotch on the rock
    D. Bourbon and water

42. Which of the following wines traditionally complements the taste of fish and seafood?
    A. Dry white
    B. Light red
    C. Hearty red
    D. Semisweet sparkling

43. The range of alcohol percentage in wines is between
    A. 3 and 6 percent.
    B. 7 and 14 percent.
    C. 15 and 30 percent.
    D. 31 and 45 percent.

44. Which of the following is the primary responsibility of a sommelier?
    A. Mixing alcoholic drinks behind the bar
    B. Developing all liquor menus
    C. Merchandising and promoting European wines to American restaurant operators
    D. Recommending and serving wines to restaurant guests

45. Which of the following wines is traditionally served at room temperature?
    A. Dry white
    B. Sweet white
    C. Sparkling
    D. Red

46. An operator marks up its wines by 60 percent. If a certain wine costs the operation $17.00, what will its selling price be (rounded to the nearest dollar)?
    A. $23
    B. $25
    C. $27
    D. $29

47. For which of the following drink orders would a bartender use a well brand?
    A. Vodka gimlet
    B. Tanqueray and tonic
    C. Absolut on the rocks
    D. B&B straight up

48. Which of the following foods contains large amounts of carbohydrates?
    A. Leafy green vegetables
    B. Grapefruit
    C. Potatoes
    D. Milk

49. A good non-meat source of protein is
    A. citrus fruits.
    B. legumes.
    C. vegetable oils.
    D. potatoes.

50. Vitamin B_1 is also known as
    A. riboflavin.
    B. niacin.
    C. thiamin.
    D. folic acid.

51. Ascorbic acid is also known as

A. vitamin A.
B. vitamin C.
C. vitamin D.
D. vitamin E.

52. Which vitamin is made in the skin when it is exposed to sunlight?

A. Vitamin A
B. Vitamin B_{12}
C. Vitamin C
D. Vitamin D

Lesson 5: The Menu and Purchasing, Production, and Service

53. In purchasing, the abbreviation ROP refers to the

A. reorder point.
B. return on profits.
C. reason of purchase.
D. reissue of products.

54. A standing order is one that is

A. placed with the supplier verbally rather than in writing.
B. discounted.
C. placed by several operations together.
D. delivered at regular intervals.

55. Reference samples are given by

A. suppliers to potential purchasers.
B. purchasers to customers.
C. chefs to production employees.
D. food producers to suppliers.

56. Bidding is a form of which of the following types of buying?

A. Informal
B. Formal
C. Cooperative
D. Blank check

57. When a supplier sells items to operators for their original cost plus a specified markup, it is an example of which of the following?

A. Blank check buying
B. Cost plus buying
C. Cooperative buying
D. Formal buying

58. Specifications indicate which of the following?

A. Where purchased items are to be delivered
B. Par stock levels and when regular orders are to be delivered
C. Which of an operation's employees are authorized to accept deliveries
D. Name, amount, brand, packaging, grade, and other characteristics of items to be purchased

59. IMPS numbers simplify the purchase of which of the following items?

A. Meat
B. Produce
C. Staples
D. Frozen prepared products

60. A grade gives purchasers an idea of a product's
    A. size.
    B. region of origin.
    C. quality.
    D. price.

61. A standard of identity is a
    A. legal description of what an item is.
    B. federal quality grade.
    C. grade given within an industry rather than by a governmental agency.
    D. nutritional label.

62. All deliveries to an operation should be weighed, counted, and checked against the original purchase order and the
    A. requisition form.
    B. supplier's invoice.
    C. call sheet.
    D. quotation sheet.

63. Mise en place refers to which of the following?
    A. The tableside cart used in some service styles
    B. The classical organization of production employees and servers
    C. Getting everything ready for the job to be done
    D. Showing a wine bottle's label and cork to the host before pouring

64. Which of the following is true of counter service?
    A. It uses less space than other types of service.
    B. Efficient work spaces for servers are essential.
    C. It suits higher menu prices.
    D. It promotes low customer seat turnover.

65. In a scramble cafeteria system, customers
    A. go to different areas to pick up different foods.
    B. do not pay for their food until after they have finished eating.
    C. are offered only breakfast items.
    D. must sit at very long communal tables.

66. Which of the following types of service generally leads to the most food waste?
    A. Counter service
    B. Cafeteria service
    C. Buffet service
    D. Seated service

67. A smorgasbord is a form of which type of service?
    A. Seated
    B. Buffet
    C. Counter
    D. Cafeteria

68. In the place setting used for American service, which of the following is placed to the left of the plate?
    A. Knife
    B. Soup spoon
    C. Water glass
    D. Dinner fork

69. In French service, the guéridon is which of the following?
    A. Tableside cart
    B. Heating unit
    C. Head server
    D. Apprentice

70. Regarding guest checks, a floater is a(n)
    A. extra check given to a server in case another is lost.
    B. check being used twice by a dishonest server.
    C. customer who leaves the operation with the check and without paying.
    D. check totaled by a computerized system.

Lesson 6: Computers and Finances in Menu Planning

71. Which of the following computer applications is considered a front-of-the-house function?
    A. Menu planning
    B. Accounting
    C. Purchasing
    D. Guest sales transactions

72. With regard to computers, the term remote means that a part of the system, such as a printer, is
    A. not connected to the computer itself.
    B. located away from the computer.
    C. capable of only limited functions.
    D. produced by a different manufacturer than the computer itself.

73. In a typical point-of-sale (P-O-S) system, employees enter orders into the system through a
    A. preset keyboard.
    B. microphone.
    C. sheet of paper fed into the computer terminal.
    D. pen written directly on the computer's screen.

74. On a P-O-S system, an open check report shows which of the following?
    A. All check totals for the day
    B. Check averages over several days
    C. Productivity of individual servers
    D. Checks that have not yet been closed out

75. Sales forecasts are most commonly based on which of the following?
    A. Sales histories
    B. Desired profits
    C. Industry reports
    D. Industry averages

76. A computerized order report is used in which of the following functions?
    A. Accounting
    B. Bookkeeping
    C. Payroll
    D. Purchasing

77. Use forecasts predict which of the following?
    A. Sales for a period
    B. Labor and materials needed for a period
    C. Profits for a period
    D. Business volume for a period

78. When a computer "explodes" a standard recipe, it does which of the following?
    A. Loses record of it in its bank of files
    B. Takes it out of the list of active files when the item has been removed from the menu
    C. Calculates new ingredient amounts based on the day's desired portion yield
    D. Prints out a hard copy

79. Which of the following pieces of information would be part of a feasibility study's market study section?
    A. Pro forma profit-and-loss statement
    B. Total investors' capital
    C. Predicted return on investment
    D. Ages and incomes of potential customers

80. An operation has $83,440 in current assets and $55,627 in current liabilities. What is its current ratio?
    A. 1:1
    B. 1.5:1
    C. 2:1
    D. 2.5:1

Practice Test Answers and Text Page References

1. C..................p. 3
2. B..................p. 5
3. C................p. 10
4. C................p. 15
5. B................p. 15
6. C................p. 22
7. A................p. 23
8. A.........pp. 24–25
9. D...............p. 30
10. C................p. 31
11. A................p. 32
12. B................p. 42
13. A................p. 45
14. D...............p. 54
15. D...............p. 54
16. B................p. 58
17. B................p. 59
18. C................p. 59
19. D...............p. 65
20. C..............p. 100
21. B..............p. 101
22. D.............p. 101
23. C..............p. 111
24. B..............p. 111
25. B..............p. 114
26. A..............p. 116
27. B..............p. 143
28. D.............p. 144
29. A..............p. 146
30. C..............p. 146
31. B..............p. 147
32. B..............p. 150
33. A..............p. 150
34. B..............p. 153
35. C.....pp. 153–154
36. B..............p. 167
37. C..............p. 167
38. A..............p. 191
39. C..............p. 195
40. B..............p. 209
41. B..............p. 211
42. A..............p. 214
43. B..............p. 220
44. D.............p. 220
45. D.............p. 221
46. C..............p. 224
47. A..............p. 226
48. C..............p. 251
49. B..............p. 252
50. C..............p. 255
51. B..............p. 256
52. D.............p. 254
53. A..............p. 267
54. D.............p. 270
55. A..............p. 270
56. B..............p. 270
57. B..............p. 271
58. D.............p. 271
59. A..............p. 272
60. C..............p. 274
61. A..............p. 276
62. B..............p. 277
63. C..............p. 293
64. B..............p. 296
65. A..............p. 297
66. C..............p. 298
67. B..............p. 298
68. D.............p. 299
69. A..............p. 301
70. B..............p. 308
71. D.............p. 314
72. B..............p. 315
73. A..............p. 315
74. D.............p. 317
75. A..............p. 319
76. D.............p. 321
77. B..............p. 324
78. C..............p. 326
79. D.............p. 336
80. B..............p. 345